Change and the 2020 Census

NOT WHETHER BUT HOW

Panel to Review the 2010 Census

Thomas M. Cook, Janet L. Norwood, and Daniel L. Cork, *Editors*

Committee on National Statistics

Division of Behavioral and Social Sciences and Education

NATIONAL RESEARCH COUNCIL
OF THE NATIONAL ACADEMIES

THE NATIONAL ACADEMIES PRESS
Washington, D.C.
www.nap.edu

THE NATIONAL ACADEMIES PRESS   500 Fifth Street, NW   Washington, DC 20001

NOTICE: The project that is the subject of this report was approved by the Governing Board of the National Research Council, whose members are drawn from the councils of the National Academy of Sciences, the National Academy of Engineering, and the Institute of Medicine. The members of the committee responsible for the report were chosen for their special competences and with regard for appropriate balance.

The project that is the subject of this report was supported by contract no. YA1323-09CN0039 between the U.S. Census Bureau and the National Academy of Sciences. Support of the work of the Committee on National Statistics is provided by a consortium of federal agencies through a grant from the National Science Foundation (No. SES-0453930). Any opinions, findings, conclusions, or recommendations expressed in this publication are those of the author(s) and do not necessarily reflect the views of the organizations or agencies that provided support for the project.

International Standard Book Number-13: 978-0-309-21121-5
International Standard Book Number-10: 0-309-21121-2

Additional copies of this report are available from the National Academies Press, 500 Fifth Street, NW, Washington, DC 20001; (202) 334-3096; Internet, http://www.nap.edu.

Copyright 2011 by the National Academy of Sciences. All rights reserved.

Printed in the United States of America

Suggested citation: National Research Council. (2011). *Change and the 2020 Census: Not Whether But How.* Panel to Review the 2010 Census. Thomas M. Cook, Janet L. Norwood, and Daniel L. Cork, eds. Committee on National Statistics, Division of Behavioral and Social Sciences and Education. Washington, DC: The National Academies Press.

THE NATIONAL ACADEMIES
Advisers to the Nation on Science, Engineering, and Medicine

The **National Academy of Sciences** is a private, nonprofit, self-perpetuating society of distinguished scholars engaged in scientific and engineering research, dedicated to the furtherance of science and technology and to their use for the general welfare. Upon the authority of the charter granted to it by the Congress in 1863, the Academy has a mandate that requires it to advise the federal government on scientific and technical matters. Dr. Ralph J. Cicerone is president of the National Academy of Sciences.

The **National Academy of Engineering** was established in 1964, under the charter of the National Academy of Sciences, as a parallel organization of outstanding engineers. It is autonomous in its administration and in the selection of its members, sharing with the National Academy of Sciences the responsibility for advising the federal government. The National Academy of Engineering also sponsors engineering programs aimed at meeting national needs, encourages education and research, and recognizes the superior achievements of engineers. Dr. Charles M. Vest is president of the National Academy of Engineering.

The **Institute of Medicine** was established in 1970 by the National Academy of Sciences to secure the services of eminent members of appropriate professions in the examination of policy matters pertaining to the health of the public. The Institute acts under the responsibility given to the National Academy of Sciences by its congressional charter to be an adviser to the federal government and, upon its own initiative, to identify issues of medical care, research, and education. Dr. Harvey V. Fineberg is president of the Institute of Medicine.

The **National Research Council** was organized by the National Academy of Sciences in 1916 to associate the broad community of science and technology with the Academy's purposes of furthering knowledge and advising the federal government. Functioning in accordance with general policies determined by the Academy, the Council has become the principal operating agency of both the National Academy of Sciences and the National Academy of Engineering in providing services to the government, the public, and the scientific and engineering communities. The Council is administered jointly by both Academies and the Institute of Medicine. Dr. Ralph J. Cicerone and Dr. Charles M. Vest are chair and vice chair, respectively, of the National Research Council.

www.national-academies.org

PANEL TO REVIEW THE 2010 CENSUS

THOMAS M. COOK (*Co-Chair*), Independent Consultant, Dallas, TX
JANET L. NORWOOD (*Co-Chair*), Independent Consultant, Washington, DC
JACK BAKER, Geospatial and Population Studies Program, University of New Mexico
WARREN BROWN, Carl Vinson Institute of Government, University of Georgia
DONALD COOKE, Esri, Redlands, CA
IVAN P. FELLEGI, Statistics Canada (emeritus), Ottawa
ARTHUR M. GEOFFRION, Anderson School of Management, University of California, Los Angeles (emeritus)
SUSAN HANSON, Graduate School of Geography, Clark University
DAVID R. HARRIS,[*] Office of the Deputy Provost and Department of Sociology, Cornell University
MICHAEL D. LARSEN, Department of Statistics and Biostatistics Center, George Washington University
GEORGE T. LIGLER, Private Consultant, Potomac, MD
NATHANIEL SCHENKER, Office of Research and Methodology, National Center for Health Statistics
JUDITH A. SELTZER, Department of Sociology, University of California, Los Angeles
C. MATTHEW SNIPP, Department of Sociology, Stanford University
JOHN H. THOMPSON, National Opinion Research Center at the University of Chicago

DANIEL L. CORK, *Study Director*
CONSTANCE F. CITRO, *Senior Program Officer*
MICHAEL L. COHEN, *Senior Program Officer*
ANTHONY S. MANN, *Program Associate*

---

[*]Resigned from the panel March 23, 2010.

COMMITTEE ON NATIONAL STATISTICS
2010–2011

LAWRENCE D. BROWN (*Chair*), Department of Statistics, The Wharton School, University of Pennsylvania
JOHN M. ABOWD, School of Industrial and Labor Relations, Cornell University
ALICIA CARRIQUIRY, Department of Statistics, Iowa State University
WILLIAM DUMOUCHEL, Oracle Corporation, Waltham, MA
V. JOSEPH HOTZ, Department of Economics, Duke University
MICHAEL HOUT, Department of Sociology, University of California, Berkeley
KAREN KAFADAR, Department of Statistics, Indiana University, Bloomington
SALLIE KELLER, Science and Technology Policy Institute, Washington, DC
LISA LYNCH, Heller School for Social Policy and Management, Brandeis University
SALLY MORTON, Department of Biostatistics, Graduate School of Public Health, University of Pittsburgh
JOSEPH NEWHOUSE, Division of Health Policy Research and Education, Harvard University
SAMUEL H. PRESTON, Population Studies Center, University of Pennsylvania
HAL STERN, Department of Statistics, University of California, Irvine
ROGER TOURANGEAU, Joint Program in Survey Methodology, University of Maryland, and Survey Research Center, University of Michigan
ALAN ZASLAVSKY, Department of Health Care Policy, Harvard Medical School

CONSTANCE F. CITRO, *Director*

Acknowledgments

The Panel to Review the 2010 Census wishes to thank the many people who have contributed to its work during its first year of operation and to this first interim report. The Census Bureau, under the leadership of director Robert Groves and deputy director Thomas Mesenbourg, has been willing and eager to engage with the panel, for which we are grateful; at the U.S. Department of Commerce, under secretary for economic affairs Rebecca Blank and deputy under secretary Nancy Potok supported the work of the panel. Kevin Deardorff very ably served as the panel's primary liaison with the Bureau for the latter half of the panel's first period of work, as did Sally Obenski in the panel's formative days. Early support from and interactions with Daniel Weinberg (assistant director for decennial census and American Community Survey), Frank Vitrano (chief, Decennial Management Division), and Tim Trainor (chief, Geography Division) were critical in getting the study started. We thank all those Census Bureau staff who contributed their time and talents to the panel's plenary meetings and activities, among them: Patrick Cantwell, Robert Colosi, Arnold Jackson, Gail Leithauser, Brian McGrath, Patricia McGuire, Adrienne Oneto, Michael Palensky, Robin Pennington, Dean Resnick, Dennis Stoudt, and Michael Thieme.

Between February and November 2010, panel members and staff conducted 58 site visits to local census offices, regional census centers, data capture centers, and other locations to observe 2010 census field operations in progress (see Appendix B for a listing). We will say more about these visits and the impressions drawn from them in our later reports but, in this first public report, we would be remiss if we did not acknowledge the temporary census field staff we encountered during these visits and thank them for their service and dedication to a difficult job, their patience in answering our every question, and their candor.

At the panel's request, 2011 census program manager Marc Hamel from Statistics Canada participated in a Census Bureau brainstorming session on response options in November 2009. In May 2010, Hamel and assistant chief statistician Peter Morrison hosted a wide-ranging briefing for a working group of the panel on Canadian census operations and the development of Internet response options in Canada. We thank them and the other talented Statistics Canada staff with whom we have spoken, and we look forward to further interactions as the 2011 census of Canada unfolds.

Arnold Greenland (IBM) and Larry Stone (Metron) accepted our invitation to participate in a brainstorming session with Census Bureau staff and a working group of panel members in November 2009 on modern operations engineering and technology insertion in the census context; we thank them for their contribution to that session. We also thank Roger Tourangeau (Joint Program in Survey Methodology, University of Maryland and University of Michigan) for his participation in two of the panel's site visits and, in so doing, helping to act as a point of contact between the panel and our parent Committee on National Statistics.

We would like to thank Constance Citro and Michael Cohen for their active participation in our panel meetings; their knowledge and expertise on census history and processes were extremely helpful during our deliberations of the panel. Anthony Mann provided timely and excellent logistical support for all our meetings and field trips and took care of the many details necessary for the panel to function effectively. Study director Daniel Cork did and continues to do an outstanding job of helping to lead the panel. He has done an extraordinary job of documenting the panel's field visits, summaries which have been instrumental in giving the entire panel the benefit of learning from every trip even if he or she was unable to make the trip. He also did a masterful job of drafting this first report and making the appropriate changes in response to the comments of the reviewers.

Finally, it has been a pleasure interacting with the other members of the panel who are extremely talented, knowledgeable, focused, and highly motivated to understand the census process in depth with the goal of improving the quality and cost effectiveness of the 2020 census.

This report has been reviewed in draft form by individuals chosen for their diverse perspectives and technical expertise, in accordance with procedures approved by the Report Review Committee of the National Research Council. The purpose of this independent review is to provide candid and critical comments that will assist the institution in making the published report as sound as possible and to ensure that the report meets institutional standards for objectivity, evidence, and responsiveness to the study charge. The review comments and draft manuscript remain confidential to protect the integrity of the deliberative process. We thank the following individuals for their participation in the review of this report: Peter Bajcsy, Image

Spatial Data Analysis Group, National Center for Supercomputing Applications, and Data Analytics and Pattern Recognition, Institute for Computing in Humanities, Arts and Social Science (I-CHASS), University of Illinois at Urbana-Champaign; Don A. Dillman, Social and Economic Sciences Research Center, Washington State University; C.A. ("Al") Irvine, Consultant, San Diego, CA; Sallie Keller, Science and Technology Policy Institute, Washington, DC; Benjamin F. King, Statistical Consultant, Durham, NC; Edward B. Perrin, Professor Emeritus, Department of Health Services, University of Washington; Robert J. Willis, Institute for Social Research, University of Michigan; and Kirk M. Wolter, Executive Vice President, National Opinion Research Center at the University of Chicago.

Although the reviewers listed above provided many constructive comments and suggestions, they were not asked to endorse the conclusions or recommendations nor did they see the final draft of the report before its release. The review of this report was overseen by Philip J. Cook, ITT/Sanford Professor of Public Policy Studies, Sanford School of Public Policy, Duke University, and Charles F. Manski, Board of Trustees Professor in Economics, Department of Economics, Northwestern University. Appointed by the National Research Council, they were responsible for making certain that an independent examination of this report was carried out in accordance with institutional procedures and that all review comments were carefully considered. Responsibility for the final content of this report rests entirely with the authoring committee and the institution.

Thomas M. Cook, *Co-Chair*
Janet L. Norwood, *Co-Chair*
Panel to Review the 2010 Census

Contents

Summary		1
Change and the 2020 Census: Not Whether But How		5
A	The Panel and This Report	7
B	Research Plans for the 2020 Census	8
C	2020 Directions: Positive Signs, But Focus and Commitment Needed	8
D	Field Reengineering: Need for Modern Operations Engineering	13
E	Response Options: Promoting Easier and Less Expensive Replies	18
F	Administrative Records: Supplementing Multiple Census Operations	20
G	Geographic Resources: Measuring Quality and Updating Continuously	24
References		29
A	Charge of the Panel to Review the 2010 Census	33
B	Site Visits by Panel Members and Staff	35
C	Biographical Sketches of Panel Members and Staff	39

Summary

SPONSORED BY THE CENSUS BUREAU and charged to evaluate the 2010 U.S. census with an eye toward suggesting research and development for the 2020 census, the Panel to Review the 2010 Census uses this first interim report to suggest general priorities for 2020 research. Although the Census Bureau has taken some useful organizational and administrative steps to prepare for 2020, the panel offers three core recommendations, by which we suggest that the Census Bureau take an assertive, aggressive approach to 2020 planning rather than casting possibilities purely as hypothetical.

The first recommendation on research and development suggests four broad topic areas for research early in the decade:

Recommendation 1: The Census Bureau should focus its research and development efforts on four priority topic areas, in order to achieve a lower cost and high-quality 2020 census:

- Field Reengineering—applying modern operations engineering to census field data collection operations to make the deployment of staff and the processing of operational data more efficient;

- Response Options—emphasizing multiple modes of response to the census for both respondent convenience and data quality, including provision for response via the Internet;

- Administrative Records—using records-based information to supplement and improve a wide variety of census operations; and

- Continuous Improvement of Geographic Resources—ensuring that the Census Bureau's geographic databases, especially its Master Address File (MAF), are continually up-to-date and not dependent on once-a-decade overhauls.

Second, we suggest that the Bureau take an aggressive, assertive posture toward research in these priority areas:

Recommendation 2: The Census Bureau should commit to implement, in the 2020 census, strategic changes in each of the four priority areas identified in Recommendation 1. The manner of implementing them should be guided by research on how each type of change may influence the trade-off between census accuracy and cost.

We think that this approach is the most effective way to build the research and evidentiary base for the 2020 census plan.

Third, we see the setting of bold goals as essential to underscoring the need for serious reengineering and building commitment to change. Accordingly, we urge the Bureau to publicly set ambitious goals regarding the cost and quality of the 2020 census:

Recommendation 3: The Census Bureau should motivate its planning and reengineering for the 2020 census by setting a clear and publicly announced goal to reduce significantly (and not just contain) the inflation-adjusted per housing unit cost relative to 2010 census totals, while limiting the extent of gross and net census errors to levels consistent with both user needs and cost targets. This should take into account both overall national coverage errors and regional variations in them.

Within each of the four topic areas listed in Recommendation 1, the report briefly sketches high-priority research projects. In terms of field reengineering, the important task is to approach census-taking with something closer to a blank-sheet approach using modern operations engineering as the focus; articulation of the logical architecture for the census would help maintain a focus on functionality and requirements for technical systems, an area in which the Census Bureau stumbled in the development for 2010. On response options, it is most essential that the Census Bureau fully and openly monitor the implementations of Internet response options in other national censuses, particularly the aggressive "wave methodology" to be used in the 2011 census of Canada. In administrative records, the important task is to complete the Bureau's planned match of 2010 census returns with its current administrative records data system, compiled from seven federal agency contributors, but—in doing so—to get beyond the question of whether an "administrative records census" (substituting records for enumeration) is feasible and instead to find ways for administrative data to supplement the whole range of census operations. Finally, with respect to improving the Census Bureau's geographic resources, steps toward processes under which state, local, and tribal governments can provide geographic updates in an easy and reliable manner is important; the critical focus of the work should be in the development of quality metrics, to finally be able to provide hard answers to questions of how good the Bureau's address lists and street-map coverages are for specific areas.

Change and the 2020 Census: Not Whether But How

Forty years ago, the U.S. Census Bureau mailed questionnaires for the 1970 census to households "in the larger metropolitan areas and some surrounding counties"—covering roughly 60 percent of the population—and asked that the households return the completed form by mail (U.S. Census Bureau, 1976:1-6). Structured in 393 local offices and coordinated by staff in 13 regional (area) offices, a large and temporary workforce of 193,000 employees was assembled for two basic but massive tasks (U.S. Census Bureau, 1976:1-52, 5-1, 5-4).[1] First, the temporary staff conducted census operations and interviewed respondents outside the dense urban areas—covering less than half of the total population but the vast majority of the land area of the nation. Second, the staff carried out the costly operation of knocking on doors and following up with households that did not return the mail questionnaire. Completed questionnaires were processed using the Bureau's Film Optical Sensing Device for Input to Computers (FOSDIC) system, preparing the data for analysis and tabulation.

The 1970 census is instructive because—in broad outlines—it has provided the basic model of U.S. census-taking for every decennial count that has followed. The numbers of offices and staff have changed in later censuses and some of the underlying technology has changed, including FOSDIC's microfilm sensing giving way to optical character recognition in 2000 and 2010. The fraction of the population counted principally by mailout/mailback of questionnaires has increased, although temporary field staff are still the first point of contact—either for delivery of questionnaires

---

[1]The number of offices does not include 6 local offices and 1 temporary regional office in Puerto Rico, and the 193,000 staffing figure does not include Alaska or Hawaii (U.S. Census Bureau, 1976:1-52, 5-4).

to be returned by mail or for direct interviewing—for people across much of the land area of the nation.

Although the methodological basics of the U.S. census have remained the same over those 40 years, the cost of the census has decidedly not. Since 1970, the per-housing-unit cost of the census increased by at least 30 percent from decade to decade (and typically more); even with the Census Bureau's announcement that the 2010 census will return $1.6 billion to the treasury, the per-housing-unit cost of the 2010 census is likely to exceed $100, relative to the comparable 1970 figure of $17 per unit (National Research Council, 2010:Table 2-2).[2] To be sure, a contributor to the cost increases has been the addition of specialized operations to increase the coverage and accuracy of the count—to the extent that the 2000 census largely curbed historical trends of undercounting some demographic groups and, indeed, may have overcounted some (National Research Council, 2010:29–30). That said, the cost of American census-taking has reached the point of being unsustainable in an era when unnecessary government spending is coming under increased scrutiny; the Census Bureau is certain to face pressure to do far better than the projections of straight-line increases in costs that have accompanied earlier censuses.

At this writing, the tremendously complex and high-stakes civic exercise that is the 2010 census will still be very much in operation. Even with the release of state-level population counts by the statutory deadline of December 31, 2010 ("within 9 months after the census date;" 13 USC §141[b]),[3] work will continue toward the release of the detailed, census-block–level data for purposes of legislative districting by the end of March 2011 (13 USC §141[c]). Indeed, even some field interviewing work for the Census Coverage Measurement (CCM) that will provide basic quality metrics for the census continues into the early months of 2011. However, although the 2010 census continues, it is not too early to turn attention to the census in 2020. It is actually both appropriate and essential that work and research begin very early in the 2010s, if the design of the 2020 census is to be more than an incremental tweak of the 2010 plan and if the 2020 census is to be more cost-effective than its predecessors.

---

[2] Figures are based on conversions to 2009 dollars in National Research Council (2010:Table 2-2). The Census Bureau announced the $1.6 billion savings (relative to budgeted totals) in August 2010; see http://www.census.gov/newsroom/releases/archives/2010_census/cb10-cn70.html. Using $13.1 billion as the life-cycle cost for the 2010 census rather than $14.7 billion as in the cited table yields a per-household-cost estimate of $102.42.

[3] The Census Bureau announced the apportionment totals 10 days early, in a December 21 press event. Earlier that day, the secretary of commerce officially transmitted the results to the president as required by law; the president, in turn, transmitted the numbers and the corresponding allocation of seats to both houses of Congress during the first day of the 112th Congress on January 5, 2011.

A THE PANEL AND THIS REPORT

Sponsored by the U.S. Census Bureau, the Panel to Review the 2010 Census has a broad charge to evaluate the methods and operations of the 2010 census with an eye toward designing and planning for a more cost-effective 2020 census. (The full statement of charge is shown in Appendix A.) In our first year of operation, the panel has held five meetings with both public data-gathering sessions and deliberative sessions. In late fall 2009, the Census Bureau convened a series of informal brainstorming sessions on possible directions for 2020—on such topics as response options and coverage improvement—in which members of our panel participated along with other external experts and Census Bureau staff. Subsequently, small working groups of panel members held similar brainstorming sessions with Census Bureau staff on topics chosen by the panel, including automation of field processes. Between February and August 2010, panel members and staff conducted 58 site visits to local census offices, regional census centers, and data capture facilities in order to obtain information on current census operations with an eye toward future improvements; see Appendix B for a listing. A subgroup of the panel also visited the headquarters of Statistics Canada in Ottawa in May 2010 to discuss the use of the Internet for data collection in Canada's 2006 and 2011 censuses, as well as the Statistics Canada approach to research and testing.

This first interim report is directed at the forward-looking part of our charge—general guidance for 2020 census planning—for two reasons alluded to in the introduction. First, there are important ways in which the 2010 census is still ongoing—neither the final census data nor the operational data needed to evaluate specific census functions are yet available—and so the actual "2010 evaluation" part of our charge is necessarily premature. More fundamentally, the second reason for a forward-looking approach is that the early years of the decade are critical to the shape of a 2020 count that is meaningfully different from 2010. Change is difficult for any large organization, and confronting the swelling cost of census-taking is something that will require aggressive planning, research, development—and, to be clear, investment of resources—in a very short time frame.

Importantly, the guidance in this report draws on the efforts and experience of several predecessor National Research Council panels. Our Panel to Review the 2010 Census effectively combines the functions of two expert panels sponsored by the Census Bureau to accompany the 2000 census: the Panel to Review the 2000 Census, tasked to observe the census in process, and the Panel on Research on Future Census Methods, tasked to evaluate the then-emerging plans for the 2010 census. Both of those panels' final reports offer recommendations for later censuses that remain relevant for 2020 planning; in particular, the suggested directions in *Reengineering the*

2010 Census: Risks and Challenges (National Research Council, 2004b) on the Census Bureau's geographic resources and its approach to developing the technical infrastructure of the census still apply directly to 2020. More recently, the immediate predecessor Panel on the Design of the 2010 Census Program of Evaluations and Experiments (CPEX) issued its final report, *Envisioning the 2020 Census* (National Research Council, 2010). Having offered its guidance on the design of the formal experiments and evaluations accompanying the 2010 census in its earlier reports, that panel's final report deliberately addressed census research and evaluation in a much broader perspective. Consequently, our panel's report serves to amplify and extend some of the themes from the CPEX panel's study.

B RESEARCH PLANS FOR THE 2020 CENSUS

At steps during the first year of our work, the panel reviewed initial suggestions by the Census Bureau for their research plan leading to the 2020 census. We have done so in our plenary meetings as well as the working group sessions mentioned above. However, we cannot comment on the Bureau's "final" version of its initial research plan for 2020 because it is not available to us. The intricacies of the federal budgeting process are such that *some* form of a research plan is factored into the Census Bureau's budget submission for fiscal year 2012. However, as those submissions are not official in any sense until the administration formally proposes its budget in early 2011, the Census Bureau is not at liberty to discuss specific proposals with us or any other advisory group.

We think it counterproductive to try to assess and speak about specifics in the research plan because we have no insight as to what details may or may not have made it into a final draft, not to mention what may or may not change the plan during departmental and administration review. Accordingly, this deliberately-short report focuses on general principles, reserving discussion of specific proposals for future reports and interactions.

C 2020 DIRECTIONS: POSITIVE SIGNS, BUT FOCUS AND COMMITMENT NEEDED

With that caveat, our general assessment of the Bureau's posture going into early 2020 census research and planning is that there are several good signs. The Census Bureau deserves credit for early moves that it has made in the direction of 2020 census planning. The Bureau's expressed intent to create a parallel organizational directorate on 2020 while the existing 2010-focused directorate continues its work and its reinstatement of a core research and methodological directorate are both very positive signs. We are also encouraged by the Bureau's apparent intent to use smaller, more frequent experiments during the decade rather than rely principally on a small number of large-scale "test censuses" as in recent decennial cycles.[4] At its December 2010 meeting, the panel also heard about the Bureau's commitment to improve its cost accounting and its analytical cost modeling capabilities, both of which will be essential to informing the discussions of census cost and quality that await the decade. Finally, we think it is a positive sign that—as we have discussed broad sketches of a 2020 research strategy with the Census Bureau over the first year of the panel's work—there appears to be agreement between the Bureau and the panel on the broad topic areas along which an effective and efficient 2020 census design should take shape.

The guidance on research offered by two predecessor National Research Council panels—one that reviewed plans for the 2010 census early in the last decade (National Research Council, 2004b) and a more recent one that looked ahead to 2020, having reviewed the experiments and evaluations planned for the 2010 census (National Research Council, 2010)—remains sound. We explicitly echo some of their points here and generally endorse their suggestions. However, we also share their concerns that early census planning efforts will founder if they lack a clear focus and strong organizational commitment. Accordingly, in this report, we are deliberately very sparing in our formal recommendations—reserving them to three main messages that are meant to suggest and cultivate a specific attitude toward 2020 census research.

First, we suggest that research and development energies be focused under four headings:

Recommendation 1: The Census Bureau should focus its research and development efforts on four priority topic areas, in order to achieve a lower cost and high-quality 2020 census:

- Field Reengineering—applying modern operations engineering to census field data collection operations to make the deployment of staff and the processing of operational data more efficient;
- Response Options—emphasizing multiple modes of response to the census for both respondent convenience and

---

[4]See National Research Council (2010:App. A) for a summary of major Census Bureau testing and experimentation activities between 1950 and 2010. From that summary, the National Research Council (2010:65–67) panel found it clear that "the Bureau used to be considerably more flexible" in its testing and that "small, targeted tests in selected sites used to be more frequent;" by comparison, in the 2000 and—particularly—2010 rounds of testing, "selected studies seem to have been chosen more based on the availability of testing 'slots' " in large-scale tests than on research questions or operational concerns. As that panel noted, there is certainly value in large-scale tests or dress rehearsals to properly practice with new census machinery, but there are many more questions that can be answered through strategic use of smaller-scale research activities.

data quality, including provision for response via the Internet;
- Administrative Records—using records-based information to supplement and improve a wide variety of census operations; and
- Continuous Improvement of Geographic Resources—ensuring that the Census Bureau's geographic databases, especially its Master Address File (MAF), are continually up-to-date and not dependent on once-a-decade overhauls.

We urge the Census Bureau to adopt a small number of focused goals. Individual research projects should be considered and conducted with reference to these priority areas; consideration should be given to how individual research efforts build on each other and contribute to an overall program of research within each topic area.

The key problems that we have observed in early iterations of the Bureau's 2020 research plan are that—beyond identifying these broad, priority topic areas—the Bureau's plans have shown a lack of focus and a lack of commitment, and they have suffered somewhat from the "stovepipe" mentality that the Bureau's new organizational approaches may help to break. It may be useful to elaborate on each of those phrases:

- By "lack of focus," we mean that initial drafts of the research plan included dozens of specific projects, roughly falling under four main topic headings in one iteration but with little notion of how they contribute to that topic and how (or if) they build from one to the other. The point in laying out a research agenda is to provide some kind of direction toward an end result, outlining how specific research tasks shed light on the decisions and trade-offs that will ultimately need to be made in shaping the 2020 census; previous versions of the Bureau's research plan appeared to try to overwhelm with the sheer number and range of activities, and lacked that sense of direction.

- By "lack of commitment," we mean that the Bureau seems to have largely shied from taking more than an exploratory position to these four priority areas—not wanting to be locked into any one design too early, which is understandable, but ultimately conveying a sort of half-heartedness about major changes in approach. The argument that "no one knows what X will look like in 2020"—in which, in varying discussions, X has been "the Internet," "mail delivery," "administrative records," "commercial records," "geography," and others—is undeniably true. But that reasoning is dangerous if it is used to dispel or minimize research on future directions rather than a challenge to work on those aspects of future technology and capability that *can* be studied and tested now—anticipating the kinds of capabilities that will be available in commercial, off-the-shelf hardware and software closer to 2020.

- Finally, by the "stovepipe" mentality, we mean that—at least until the recent administrative changes began to circulate—the 2020 planning was being done on scarce resources by very limited numbers of staff. The staff work was energetic and good and very useful for framing, but ultimately lacking because it was—and was presented as—a set of proposed activities done in isolation from other parts of the Bureau. Among other things, the draft research proposals for 2020 lacked any explicit connection or coordination with the formal research work being done in the 2010 CPEX program, the activities planned in the Geographic Support System initiative (about which more is said below), the American Community Survey (including the testing of Internet response to that survey) and other current surveys, and the Bureau's economic directorate (which also makes use of Internet response and operational control systems).

Continuing to ask *whether* the Bureau should retool its field technical infrastructure or *whether* administrative records should play a role in the 2020 census is not the right approach; it seems to be grounded in the notion that a single fix or a single tweak in census approach will be sufficient to drive down 2020 census costs. We are convinced that no such single fix exists, and that the shape of the 2020 census will have to make use of work in *all* the priority research areas, in some measure, to materially change 2020 conduct. Accordingly, we recommend an aggressive, assertive posture toward research:

Recommendation 2: **The Census Bureau should commit to implement, in the 2020 census, strategic changes in each of the four priority areas identified in Recommendation 1. The manner of implementing them should be guided by research on how each type of change may influence the trade-off between census accuracy and cost.**

We think that this approach is the most effective way to build the research and evidentiary base for the 2020 census plan.

The third and final central message of this report is meant as a practical way of underscoring and cultivating the kind of commitment to serious reengineering called for in Recommendation 2. We concur with the predecessor National Research Council (2010:43) panel that commitment to change can and should be helped by setting a bold goal that is "stark, ambitious, and public." As the previous panel wrote, "it has become almost rote to include 'containing cost' as a goal of the decennial census" when what is needed is "meaningful reductions in per-household cost—through leveraging new technology and methodology—without impairing quality."

We agree that a bold goal is crucial to motivating census research over the decade, and accordingly suggest a slight variant of the goal offered by the previous panel:

> *Recommendation 3:* **The Census Bureau should motivate its planning and reengineering for the 2020 census by setting a clear and publicly announced goal to reduce significantly (and not just contain) the inflation-adjusted per housing unit cost relative to 2010 census totals, while limiting the extent of gross and net census errors to levels consistent with both user needs and cost targets. This should take into account both overall national coverage errors and regional variations in them.**

Quite deliberately, we phrase our recommendation in still-stark but more general terms than the previous panel, which urged (National Research Council, 2010:Rec. 2.1) the Bureau to plan for the 2020 census with the stated goal of holding per-household cost and national and major demographic group coverage errors to their *2000* census levels (not 2010). The previous panel's report (National Research Council, 2010:Ch. 2) traces long-term trends in census cost and quality measures, noting the more than 600 percent increase in real-dollar per-household cost between the 1970 and 2010 censuses in contrast with much smaller relative gains in census accuracy (as measured by net census error). An earlier National Research Council (1995:55) panel devoted considerable attention to explaining the growth in census cost between the 1970 and 1990 censuses, finding itself unable to directly account for some three-fourths of the total increase. That panel ultimately concluded that the increase was largely driven by the Census Bureau "pouring on resources in highly labor-intensive enumeration efforts to count every last person," in response to demands for highly accurate small-area data, at the same time as public cooperation with the census dipped and measured net undercount actually increased from 1980 to 1990. Fifteen years later, the successor National Research Council (2010:39–40) concurred, noting a "steady accretion of coverage improvement programs" over the decades—all of which arguably have some value but few of which are subjected to extensive cost-benefit analysis, and none of which are cost-free. That panel observed that, looking ahead to 2020, the census is at a critical point at which additional spending on existing methods or adding still more coverage improvement layers "in an effort to reduce net undercoverage could conceivably add more error than it removes." We think that the research directions we suggest in this report are capable of achieving significant streamlining of effort and per-capita household cost reductions without tipping the balance to higher levels of census error. But we also think that it is premature to suggest specific totals or percentages as targets for 2020; setting those targets will depend critically on the raw and operational data from the 2010 census, the results of the 2010 census evaluation and Census Coverage Measurement programs—and on early, pilot research this decade.

D FIELD REENGINEERING: NEED FOR MODERN OPERATIONS ENGINEERING

The priority research areas noted in Recommendation 1 are all important, but it is logical to start the discussion with the topic of field reengineering and the automation of field operations. We use the term "field reengineering" as a convenient shorthand, cognizant that the term is open to overly simplistic interpretations—"field" perhaps connoting a narrow focus on the moment-by-moment work of individual temporary enumerators and local staff, "reengineering" perhaps connoting a stringent restriction to development of computer software or hardware systems. By the term "field reengineering," we mean both of those individual threads and more: a fundamental evaluation of all major operations, with an eye toward optimization of effort and resources and improvement of cost-effectiveness.

If cost reduction—while maintaining quality—is to be a major focus of planning the 2020 census, then it follows that reexamining and streamlining field operations must be on the table. The largest component expenses of the modern decennial census are those that involve the mass hiring and deployment of temporary census workers. The most expensive single operation in modern censuses is Nonresponse Follow-up (NRFU), knocking on doors and otherwise attempting to collect information from households that do not return their questionnaire by mail or other means. Doing whatever is possible to contain the size of the NRFU workload is a key motivator for work on administrative records and response options, as discussed below. But, assuming that there will inevitably be a need for some substantial NRFU, finding ways to make it more cost-effective is important. Although NRFU is the single largest field operation, other major field operations also involve the major deployment of temporary staff: in support of the 2010 census, such operations included the complete Address Canvassing operation to verify address list entries in 2009, a series of operations to establish contact with and then count at group quarters (nonhousehold) locations, and the deployment of enumerators to either deposit questionnaires or conduct interviews in areas of the country where mail delivery of questionnaires was not feasible.

It is also appropriate to discuss field reengineering and automation first because they may be the most difficult for the Census Bureau to address, on three key levels. One is that true, systematic review of operations—close to approaching the basic ideas of census-taking from a blank-sheet or first principles approach—is relatively new to the decennial census. The established and entrenched mechanics of census-taking, stemming from having used the same basic outline of operations for 40 years, breeds a familiarity with normal routines; this familiarity, in turn, contributes to a culture in which "just-in-time" systems development and training are accepted, even though the risks are high and costs substantial. A second reason for the primacy of field reengineering is that it is cross-cutting and highly intertwined with the other three research areas. The technical systems that assign field staff must properly synchronize with systems for handling multiple response modes to the census form, the degree to which administrative records data may be used in census operations directly affects the scope of field operations and the level of follow-up necessary, and field systems are of little use if they do not reflect current and accurate geographic features and address information. Accordingly, a fresh approach to field automation can be difficult because the task is so large and extensive, and so it is not neatly compartmentalized into a single "project."

But, arguably, the key difficulty in field reengineering is illustrated by the record of experience leading to the *2010* census. A more detailed account of systems development for 2010 must await future reports of the panel, in line with evaluating the systems that ultimately were used in the census. But for this first report a brief summary suffices. The complication for approaching field reengineering in 2020 is that field automation—in the specific form of developing handheld computers for use in both NRFU and Address Canvassing—was a major plank in the Census Bureau's plans for 2010. The Census Bureau assembled mobile computers using commercial, off-the-shelf products for preliminary testing in its 2004 and 2006 field tests. As those tests continued, the Bureau moved toward issuance of a major contract to develop not only the handheld computers but also the operational control systems that manage the flow of information among census headquarters, regional and local census offices, and individual field staff. The five-year, $600 million Field Data Collection Automation (FDCA) contract was awarded to Harris Corporation in March 2006. Problems with use of the Harris-developed devices in spring and summer 2007, in Address Canvassing for the 2008 census dress rehearsal, began to be noted during that operation (e.g., U.S. Government Accountability Office, 2007) and into the fall. In early January 2008, online media broke the story that the Census Bureau had been advised in November 2007 that the handheld development was in sufficiently "serious trouble" that paper-based contingency operations should be immediately developed (Holmes, 2008). When the Census Bureau submitted a new, "final" set of requirements to its FDCA contractor in mid-January 2008, the resulting cost estimate prompted the Bureau and the U.S. Department of Commerce to assemble a task force to suggest options for the

FDCA work.[5] Ultimately, the strategy chosen by the Bureau in 2008 was to abandon the use and the development of the handhelds for all but the Address Canvassing operation—making the 2010 NRFU operation completely paper-based and putting the estimated total life-cycle cost of the 2010 census at roughly $14.5 billion.[6]

That the Census Bureau stumbled in field systems development—very visibly and expensively—in preparing for the 2010 census is a complication for 2020 because it may induce some skittishness about moving aggressively so early in the development for the next census. The high price tag of the collapse of the FDCA handhelds and the late switch to paper-based NRFU operations may also make it more difficult to sell the idea of field reengineering as a short-term investment to save money with an efficient and effective census in the long run. But we think it is a wise investment, and that it is key to avoiding a 2020 census that is merely an incremental tweak on 2010; having stumbled in 2010 systems development highlights the importance of trying again and succeeding where the previous efforts foundered.

Our predecessor National Research Council (2004b:172–173) Panel on Research on Future Census Methods sketched out the major stages of successful system reengineering efforts, based on "past experience with reengineering and upgrading information technology operations within corporations and government agencies." In our assessment, these steps remain the right prescription, and we echo and endorse them here:

1. *Define a "logical architecture" or "business process" model:* A "logical architecture" is a blueprint of the workflow and information flow of a particular enterprise—the full set of activities and functions and the informational dependencies that link them. As described by the earlier panel (National Research Council, 2004b:175), the key attribute of the logical architecture model is that it is focused on function and purpose; it is not a timetable that assigns completion times to individual functions and it should not be based on (or constrained by) existing organizational boundaries. The baseline logical architecture model becomes just that: an "as-was" model that serves as the basis for redesign or replacement.

---

[5] Our predecessor Panel on Research on Future Census Methods—whose final report, issued in 2004, we reference in this section—forecast the problems that burst forth in 2008, stating that the handheld development effort would go awry without early attention to requirements and functionality rather than specific forms of devices. "A second risk inherent with the [handheld computer] technology lies in making the decision to purchase too early and without fully specified requirements, resulting in the possible selection of obsolete or inadequate devices" (National Research Council, 2004b:7).

[6] For additional information on the handheld development portion of the FDCA contracts, (see, e.g., U.S. Department of Commerce, Office of Inspector General, 2006; U.S. Government Accountability Office, 2008).

2. *Reengineer the logical architecture:* The real value of logical architecture models comes in actually using the models, reexamining assumptions and features in the "as-was" model to construct one or more "to-be" models. Changes in the to-be models can be relatively minor in isolation (i.e., correcting evident redundancies in information flows in the baseline model) or they can be sweeping, wholesale replacement—the point being that the modeling framework provides a means for ensuring that changes in part of the model fit well into the enterprise as a whole and satisfy downstream operational needs. These new candidate models can then be compared with each other and evaluated for their feasibility and efficiency in structuring functions to meet new operational demands.

3. *Construct the physical technical infrastructure using the reengineered logical architecture:* Neither the as-was architecture model nor any of the candidate to-be models are, or should be, exact drafts of finished computer hardware or software systems. That is, the logical architecture models are not directly equivalent to the finished physical, technical architecture of a particular enterprise. However, the logical architecture models do refine the *requirements* of the finished technical systems—the objectives and information demands that the final architecture must accommodate.

At the time of the earlier panel's work in the early 2000s, the Census Bureau was conducting a "reengineering exercise" and developing an as-was model based on the information flows of the 2000 census; the panel indicated that it "enthusiastically endorses and supports" this pilot work and urged its continuation, particularly to the extent that it served to document and assess the complete, end-to-end census process instead of limited parts (National Research Council, 2004b:176–177). While complimentary about the pilot work, the panel strongly cautioned that the effort would founder without strong institutional commitment. Noting that reengineering efforts can go awry without top-level management "champions," the National Research Council (2004b:Rec. 6.1) panel recommended that "the highest management levels of the Census Bureau should commit to the design and testing of a redesigned logical architecture," conveying the importance of the task and facilitating "buy in" by all the organizational divisions in the Bureau. The National Research Council (2004b:Rec. 6.2) panel further suggested the strong need to "create and staff the position of system architect for the decennial census," equipping that architect with "authority to work with and coordinate efforts" among the Bureau's divisions. Neither of these recommendations was embraced by the Bureau and, predictably, the architecture-building effort stopped well short of its potential. Though "Version 1.1" of a suggested to-be 2010 census architecture was included in the online library of reference documents for the FDCA program (http://www.census.gov/procur/www/fdca/library.html), the give-and-take between competing to-be models and the selection from them does not appear to have happened and—as mentioned above—FDCA requirements were only solidified at a too-late stage in early 2008, when they would have been more central if a reengineered logical architecture had been properly used as a template.

As the Bureau proceeds with planning the 2020 census, there is an urgent need to document the costs and benefits of *every* constituent census operation; development of a business process or architecture model should be viewed as a complementary part of that documentation, specifying the information required by and provided by each operation. Efforts could go awry again in this decade if architecture modeling is perceived as simply moving boxes around on a piece of paper—not embraced as a means of comparing and contrasting alternate approaches in a low-risk environment (relative to cobbling together prototype procedures and systems) while keeping a focus on the requirements for eventual, finished technical systems.

Consonant with our suggested "not whether but how" stance in Recommendation 2, we reiterate the previous panel's language on the pressing need for organizational, corporate commitment within the Census Bureau to a mature systems reengineering process for 2020. From a research standpoint, perhaps the most important single project that could be undertaken is articulation of the as-was model of information flows in the 2010 census and active exploration of possible to-be scenarios. Much more than just boxes and lines in a diagram, this is sophisticated work that should be undertaken with real training and mentorship by experienced practitioners of operations research and modern operations management; the Bureau's fledgling effort in architecture modeling in the early 2000s appeared to founder because it stopped short of analysis and did not decompose activities to the appropriate level to enable real reengineering (National Research Council, 2004b:49). Though the primary focus of the reengineering should appropriately be on the support systems for the decennial census, it is also important that the effort connect with (and help modify, as appropriate) the comprehensive systems of the Census Bureau, including the ongoing American Community Survey (ACS), the Bureau's other current demographic surveys, its geographic support systems, and its economic census programs as well as the decennial—facilitating ways for feedback or technical improvements in one part of the Bureau to improve the others. The eventual goals of field reengineering are systems that can efficiently handle the information and processing needs of the decennial census and its managers, tools for the efficient acquisition of raw data, and technical solutions that are easy to use and comprehend by the large corps of temporary and relatively untrained enumerators and field staff. Yet the underlying processes cannot be reengineered if they are not articulated and assessed up front.

E RESPONSE OPTIONS: PROMOTING EASIER AND LESS EXPENSIVE REPLIES

Field reengineering may be the most difficult of the topic areas for the Census Bureau to handle for a variety of reasons, but arguably the topic area for which a timidity in approach could be most costly is that of response options—and census response via the Internet in particular. Guiding respondents to submit their census information in an inexpensive and computer-ready format is critical to curbing the cost of moving and processing paper. To be sure, obtaining a substantial percentage of respondent take-up via the Internet is a challenging task; among other things, care must be taken to make sure that questions asked via paper or the Internet (or through other modes) share common structures and yield the same information regardless of mode, and the Census Bureau should tap the developing literature on building Internet participation in the census and survey contexts. But Internet response must not be treated as a far-off or unobtainable goal either, because it is likely a key contributor to a more cost-effective census.

Again, a full examination of the decisions made for the 2010 census awaits future (and more evaluative) reports, but a brief summary is useful here. Our predecessor National Research Council (2010) panel discussed the chronology in more detail in an appendix to its interim report (which is, in turn, reprinted as part of the 2010 volume). Although the 2000 census included an Internet response option (albeit an unpublicized one) and Internet response was advocated in very early planning documents for 2010, the Census Bureau reversed course in mid-decade and announced in summer 2006 that online response would not be permitted in the 2010 census. The primary arguments cited by then-Census Bureau director Louis Kincannon (2006) and a Bureau-commissioned report by the MITRE Corporation (2007) included intense worries about security (e.g., a "phishing" site set up to resemble the census) that could negatively impact the response rate as well as concerns from pilot testing that offering Internet response as an option did not significantly increase overall response rates. Acknowledging the Bureau's stance, the previous panel pointedly remarked that "the panel does not second-guess that decision, but we think that it is essential to have a full and rigorous test of Internet methodologies in the 2010 CPEX" (National Research Council, 2010:206). No such major test was included in the formal experiments of the 2010 census, although the Bureau did announce a small-scale "Internet measurement re-interview study, focused on how differently people answer questions on a web instrument from a paper questionnaire" (Groves, 2009:7) as a late CPEX addition.

Based on initial discussions of 2020 planning, we are heartened by some signs of commitment on the Bureau's part to exploring Internet response, both in preliminary testing for 2020 as well as in regular response to the American Community Survey. However, as with field reengineering, we suggest that response modes are another area in which top-level commitment and championship and commitment are critical to success. It is particularly important that the Census Bureau not continually fall back on arguments along the lines of "no one knows what the Internet will look like in 2020"—certainly a true statement, but one that misses the broader point. The argument would be on point if the goal were the polished implementation of a census questionnaire in any particular computer markup language or on any specific computer platform—but it is not. Rather, the goal is to investigate important factors that are not bound to specific platforms: mode effects of response (e.g., whether different demographic groups respond differently or more quickly to an electronic questionnaire than a paper version); the effectiveness of various cues or prompts to encourage respondents to adopt particular response modes; and the emergence of standards for security of online Internet transactions.

Decennial census planners should pursue research projects that elucidate mode effects and that guide respondents to use lower processing–cost response options, such as online response. But—importantly—they should not go into them trying to reinvent the wheel. Arguably, the most important, immediate research and development task in this area is to track and learn from the experiences of other countries that have implemented online response in the 2010 round of censuses. In particular, the soon-to-unfold case study of the 2011 Canadian census is a vital one. Conducted every five years by Statistics Canada, the Canadian census permitted online response in 2006 and achieved roughly a 20 percent Internet response rate—including considerably higher-than-anticipated Internet take-up rates in more rural provinces, where planners had not expected heavy Internet saturation (National Research Council, 2010:294–295). Statistics Canada hopes to double the Internet take-up rate in 2011; to do so, it is using a very aggressive "wave methodology" approach, under which most Canadian households will not receive a mailed questionnaire as their first contact by the census. Instead, some 60 percent of Canadian households—in areas where models suggest a high probability of Internet response—will receive only a letter (with a URL and Internet response instructions, including an ID keyed to the household address) in the first wave. Indeed, in subsequent reminder waves of contact, at least one more letter, telephone prompt, or postcard (generic, without Internet log-in information) will be tried before paper questionnaires are mailed en masse. The initial mailings (letters and postcards) will include a telephone number so that households can request a paper questionnaire, if desired. The other 40 percent of Canadian households are roughly evenly divided into two groups, one that will receive the census questionnaire as the initial mailing and the other (in more rural locations) where conventional questionnaire drop-off by enumerators will be performed. Côté and Laroche (2009) provide a basic overview of the 2006 response option and the plan for 2011.

The Canadian experience with strongly "pushing" some response options will be a useful one for the U.S. Census Bureau to monitor closely. Likewise, the Internet take-up rates and approaches to mobilizing online response in other national censuses, such as the 2011 United Kingdom census will merit examination. In addition to examining the results of the limited Internet reinterview study that was added into the 2010 CPEX program, the Census Bureau should also actively use the ACS as a testbed for decennial census methods. The ACS is already a multimode survey (mail, phone, and personal interview), and 2020 census planners should look at the emerging online response option for the ACS for guidance on how to best use and promote response modes in the decennial census. Clearly, the ACS is a more complex survey instrument than the short-form census, but the national scale of the ACS, its use of multiple data collection modes, and its overlap of census-short-form content make it an important tool for census testing, including the insertion of questions in a "methods panel" portion of the ACS sample. Experience from elsewhere in the Census Bureau is also useful to study (Internet response is permitted to the Bureau's economic censuses and surveys) and port over, as appropriate to the decennial census context.

F ADMINISTRATIVE RECORDS: SUPPLEMENTING MULTIPLE CENSUS OPERATIONS

A significant wild card in planning for the 2020 census is the potential role of administrative data—records compiled from other agencies of federal, state, tribal, or local governments, as well as records available from commercial sources. In this research area, the main challenge for census planners is building a business case for the use of such records in a wide variety of census operations—and thus overcoming some historical expectations for the role of administrative data in the census—to permit informed decisions about the extent of records usage in 2020.

To support the Administrative Records Experiment 2000 (AREX 2000) that accompanied the 2000 census—the Census Bureau's first foray into the use of records in a major census test—the Bureau constructed the first incarnation of its Statistical Administrative Records System (StARS) database using 1999-vintage data from federal agencies. A major challenge of AREX 2000 was the assembly, linkage, and unduplication of the StARS data, and a major focus of the experiment was to consider the potential utility of administrative records as a replacement for the census process—to wit, the concept of an "administrative records census" that has historically driven consideration of the topic. To that end, AREX 2000 zeroed in for detailed comparison of StARS and census data on two sites (Baltimore City and County, Maryland, and Douglas, El Paso, and Jefferson Counties, Colorado). The results of the AREX 2000 are summarized by Judson and Bye (2003) and related evaluation reports.

Having successfully built StARS, the Census Bureau decided to continue the work, formally posting notice (pursuant to requirements of the Privacy Act of 1974, 5 USC § 552a) of the establishment of StARS in a January 2000 *Federal Register* notice (65 FR 3203). The original notice indicated that StARS "will contain personally identifiable information from six national administrative federal programs" obtained from six federal agencies—"the Internal Revenue Service, Social Security Administration, Health Care Financing Administration, Selective Service System, Department of Housing and Urban Development, and the Indian Health Services." The notice also suggested that "compatible data may also be sought from selected state agencies, if available." Current incarnations of StARS rely on seven data sources from federal government agencies; the most prominent of the underlying sources is person-level information extracted from Internal Revenue Service (IRS) returns. Recent revisions and amendments to the regulatory notice in March 2009 (74 FR 12384) and October 2010 (75 FR 66061) have suggested an eventual wider scope for StARS, with the October 2010 notice indicating intent to obtain administrative record files from eight cabinet-level departments of the federal government and four other agencies, while "comparable data may also be sought from State agencies and commercial sources."[7]

To date, an important characteristic of the Bureau's StARS database is that it does not exist as a "living," ongoing entity. Rather, it is currently rebuilt anew each year, using new vintages of the underlying source data files that are intended to match the March/April reference time of the decennial census to the greatest extent possible. Consequently, year-to-year dynamics in the database are as-yet unexplored (save for comparison of the aggregate record counts to see how "close" in size the compiled StARS database is relative to the Census Bureau's national-level intercensal population esti-

---

[7] Specifically, the sources named in the notice are "agencies including, the Departments of Agriculture, Education, Health and Human Services, Homeland Security, Housing and Urban Development, Labor, Treasury, Veterans Affairs, and from the Office of Personnel Management, the Social Security Administration, the Selective Service System, and the U.S. Postal Service" (75 FR 66062).

mates). As an ongoing research enterprise, work on a continuous administrative records file would be extremely useful—permitting study of evolution of the records over time and shedding light on undercoverage and overcoverage of persons and households in both the census and the administrative data themselves. However, given the orientation of this report toward important first steps, our discussion below assumes work within the current StARS framework of regular rebuilding.

As of late 2009, the Bureau plans to conduct a full matching study of the 2010 census results to the StARS database, as an addition to the CPEX program. When first announced, the study was characterized as "mount[ing] a post-hoc administrative records census, using administrative records available to the Census Bureau" (Groves, 2009:7). Later, the concept of the study was suggested in budget submissions for fiscal year 2011; one of several initial projects in an administrative data initiative throughout the federal statistical system is "using administrative records to simulate the 2010 Census in order to thoroughly examine and document the coverage and quality of major governmental and commercial administrative record sets" (U.S. Office of Management and Budget, 2010:317). In its recent reactivation of a research and methodology directorate, the Bureau has also signaled an intent to make study of administrative data a high priority, creating a new office for administrative records research within the research directorate.

If executed fully, the proposed StARS–2010 census matching study—and ongoing Census Bureau research on administrative data quality and uses—is much more than an ambitious scaling-up of the AREX 2000 work. The study is the critical research activity in the area of administrative records and should be a critical proving ground, and we enthusiastically support its continuance. However, the key point that we make in this area—consistent with our "not whether but how" guidance—is that the Bureau resist the temptation to stop at the question of national-level coverage of StARS relative to the census. The question of whether a complete "administrative records census" is possible—as a replacement for the census—is an interesting one but has too often been the beginning and the end of discussions; that question is no longer the most important one (if it ever was), nor is it (arguably) the most interesting. We encourage the Bureau to be open to and use its matched records–census files, to explore the use of administrative data in a *supplementary* role to a wide variety of census operations. In particular, roles for administrative data as a supplementary resource to NRFU operations should be explored; as we discuss in the next section, work with administrative records should also be a key part in assessing and upgrading the Bureau's geographic resources—whether as a source of address updates throughout the decade or a way to identify areas that may require more (or less) intensive precensus address canvassing.

However imprecise they may be—and caveated as the results may need to be—the Bureau should also match StARS or other administrative data to the operational data from the 2010 census. Doing so would finally move toward empirical answers to important questions about possible roles of administrative data: for instance, whether administrative data might be a recourse to reduce the number of visits made by enumerators during NRFU (e.g., resorting to the records data after three contact attempts rather than six) or how they compare to data obtained from proxy responders, such as neighbors or landlords of absent householders. Use of administrative data in a true simulation of the 2010 count, and so compared with time-stamped data on the return of census questionnaires, may also suggest diagnostic measures that could be supplied to local and regional census managers to target staff and resources during field operations. Matched administrative records and census data would also facilitate necessary study of data quality from both sources, including the accuracy of race and Hispanic-origin data from administrative data and the degree of correspondence of "household" membership in administrative data (persons affiliated with a particular address) with the census "usual residence" concept.

Work on the administrative records matching study should contribute to the development of a business case for wider access to and use of administrative data, to inform final decisions on the use of the data. This business case includes both a utility side—a pure cost–benefit articulation—and an acceptability side. On the utility side, the administrative records simulation should permit cost modeling, for instance on the potential cost impact of resorting to records at different phases of NRFU. It also speaks to the quality of the data; a study that stops at coarse or large demographic-group measures and does not investigate the quality of records data on characteristics (rather than just counts) would be unfortunate. The utility side of the business case is arguably more important for early research work than the acceptability side, but the acceptability side must also be addressed. By the acceptability side, we mean studying whether the respondent public, census stakeholders, and Congress (as the ultimate source of direction for conducting the census) will accept the wider use of administrative data in the census (and for which purposes). This includes assessing general public sensitivity to providing private information (e.g., the housing tenure question in the 2010 census on whether a home is rented or owned, free-and-clear or with a mortgage) in a census or survey context compared to drawing that information from records sources. It also includes respondents' reactions to specific questionnaire or mailing package wording and cues that suggest the risks or benefits of comparing census returns with other data sources. From a technical standpoint, it also means documenting the effectiveness of the Bureau's data handling standards. At present, a critical part in StARS assembly is replacement of true personalized identifiers, like Social Security numbers, with a generalized Protected Identification Key (PIK); contracting with external users to deliberately try to "break" the Bureau's identifiability safeguards (and correcting any detected shortcomings) would bolster the security case.

A clear concern moving forward in administrative records work is refining the mix of data sources that are compiled and combined into a StARS-like database. The current StARS relies heavily on IRS tax data. The IRS data may be very good in terms of coverage, but the use of those data necessarily raises logistical and operational concerns, including potential impacts on response and goodwill toward the census based on being associated with tax authorities[8] as well as the regulatory clashes between the privacy protections nested in Titles 13 (Census) and 26 (Internal Revenue) of the U.S. Code. To that end, the Bureau should complete work that it has started on outlining a complete matrix of possible data sources for StARS, including state and local government resources as well as commercial files. The cost and quality (for generating data on characteristics) of the current federal-level StARS relative to one or more non-IRS StARS-alternatives should be examined in detail. The Census Bureau should also consider the quality and accessibility of data from sources beyond federal agency contributors as they pertain to the group quarters (GQ) population—people living in such places as college dormitories, correctional facilities, health care facilities, and military installations. The concept of a GQ-focused StARS built from facility or institutional records should be explored as a supplement to the traditional collection of data through distribution of questionnaires at large GQ facilities.

G GEOGRAPHIC RESOURCES: MEASURING QUALITY AND UPDATING CONTINUOUSLY

As one of our predecessor National Research Council (2004b:57) panels observed, "a decennial census is fundamentally an exercise in geography"— its core constitutional mandate is to realign the nation's electoral geography and its final data spotlight the nation's civic geography, describing "how and where the American public lives and how the characteristics of small geographic areas and population groups have changed with time." Accordingly, another National Research Council (2004a:57) panel concluded, without exaggeration, that the quality of the Census Bureau's geographic resources—

---

[8] On the significance of these concerns, as in the development of Internet response options, comparison of experience with other national statistics offices—particularly Canada—could be instructive. The 2006 Canadian census long-form sample adopted the approach of other Statistics Canada surveys of letting respondents check a box to permit Statistics Canada to use income tax returns to fill in questions on income. In all, 82.4 percent of long-form respondents chose the tax option (Statistics Canada, 2008:9), with no deleterious effects.

in particular, the accuracy of its address list—"may be the most important factor in determining the overall accuracy of a decennial census." This will continue to be true of the 2020 census, regardless of its eventual shape— any operational or methodological improvements are ultimately for naught if census data cannot be accurately linked to specific geographic locations and cross-checked and tabulated accurately.

In the 2000s, the Census Bureau undertook an eight-year MAF/TIGER Enhancements Program (MTEP), intended to address both the Bureau's Master Address File (MAF) and its Topologically Integrated Geographic Encoding and Referencing System (TIGER) geographic database. The centerpiece activity of MTEP, in turn, was the MAF/TIGER Accuracy Improvement Project (MTAIP)—a major contract issued to Harris Corporation (which later won the FDCA contract, described above) in June 2002 to realign the county-level TIGER files to improve the locational accuracy of streets and other features. Although revolutionary when developed in the mid-1980s, both the database structure and the point, line, and polygon quality in the TIGER files had become dated by the 2000 census. Our predecessor (National Research Council, 2004b:84) panel on 2010 census planning strongly echoed the need for an overhaul of TIGER but cautioned that the MTEP— nominally to improve *both* of the Bureau's core geographic resources in MAF and TIGER—had an unmistakably "TIGER-centric feel," with other components of the MTEP "seem[ing] to speak to the MAF largely as it inherits its quality from TIGER" and not materially improving the MAF in its own right. That panel took strong exception with language in extant Census Bureau planning documents for 2010 that signaled an intent to wait for a complete Address Canvassing operation in 2009 to seriously work on improving MAF quality (aside from periodic updates from Postal Service data; see National Research Council, 2004b:88). That panel also expressed concern that some supporting and later-stage objectives of the MTEP were ill specified or unspecified—among them the MTEP objective on quality metrics to document the quality of MAF/TIGER information and identify areas in need of corrective action (National Research Council, 2004b:77). Ultimately, the Bureau proceeded with a complete Address Canvassing operation—sending field enumerators to every block to verify address information and collect geographic operations, in the one 2010 census operation that was able to make use of handheld computers.

In the early drive toward the 2020 census, the Census Bureau has expressed its intent to make upgrades to its geographic resources a strong early focus. Typically, the "geographic support" account in the Census Bureau's budget includes regular maintenance of the main components of the Census Bureau's Geographic Support System (GSS): the MAF and TIGER databases. These maintenance activities include the regular updates of the MAF through the U.S. Postal Service's Delivery Sequence Files (DSF) and the annual Boundary and Annexation Survey that gathers and updates boundary information for local governments (and changes in their legal status). In its fiscal year 2011 budget request (U.S. Census Bureau, 2010), the Bureau seeks an additional $26.3 million (over the base $42.3 million request) in its geographic support account to kick off what has been dubbed the GSS initiative. The budget request summarizes the initiative simply (U.S. Census Bureau, 2010:CEN-191):

> The [initiative] supports improved address coverage, continual updating of positionally accurate road and other related spatial data, and enhanced quality measures of ongoing geographic programs. By focusing on activities that improve the [MAF] while maintaining and enhancing the spatial infrastructure that makes census and survey work possible, this initiative represents the next phase of geographic support after the MAF/TIGER Enhancement Program (MTEP).

Census Bureau staff also described the initiative at the panel's March 2010 meeting.

Consistent with our predecessor National Research Council (2004b) panel, we generally support the aims of the Bureau's GSS initiative; because the previous decade's development work was heavily TIGER-centric, we think it appropriate that the Bureau take a more balanced, close-to-MAF-centric posture to its geographic resources leading up to 2020. In particular, we welcome the expressed indication of moving toward continuous improvement of geographic resources over the whole decade, rather than gambling too heavily on one-time operations like the 2009 Address Canvassing round or the comprehensive mid-2000s TIGER realignment work.

That said, we support the Bureau's GSS Initiative work with a significant catch—the Bureau's geographic work early in the decade should include serious attention to quality metrics for *both* MAF and TIGER. The quality metrics and evaluation plank of the previous decade's MTEP slate never really materialized, and assertions that the Bureau's MAF represents a gold standard among address lists are no longer adequate or compelling. An important part of continuous improvement is being able to provide some manner of hard, quantitative information on how good MAF and TIGER are at any particular moment. Work in this area should include regular fieldwork, perhaps making use of the Bureau's ongoing corps of interviewers who collect information for regular demographic surveys, and may include the systematic collection of GPS-accurate map spot and line feature readings for comparison with TIGER and comparison of MAF and TIGER with comparable information from commercial and other sources (e.g., utility records or local conversions of rural route and other addresses to conventional city-style addresses for 9-1-1 location).

In general, we suggest that an important research priority for the Census Bureau as it exits the 2010 census is to aggressively mine and probe its current MAF. This includes ties to 2010 census operational data—the Bureau's knowledge of information added in the full Address Canvassing operation and late field operations like the Vacant/Delete Check as well as its knowledge of census mailings returned as "undeliverable as addressed." An earlier National Research Council (2004a) panel noted that evaluating the MAF and suggesting operational improvements were severely complicated because the structure of the Bureau's geographic sources did not readily allow for the unique contributions of individual operations (e.g., the Local Update of Census Addresses returns suggested by local and tribal governments or the regular refreshes from Postal Service data) to be disentangled and compared. Ideally, the 2020 MAF/TIGER structure is more amenable to reconstructing such operational histories for individual addresses or street features; accurate cost-benefit assessment of geographic support operations for 2020 depends vitally on the collection and analysis of these kinds of metadata.

The importance of vigorous, intensive analysis of the quality of MAF/TIGER cannot be overstated. It is tempting, but misguided, to minimize such work as simply clerical or as an exercise in fine-tuning cartographic accuracy. Spatial data quality is inextricably linked to census quality and, to the greatest extent possible, both the spatial data in MAF/TIGER and census operational data demand study at fine-grained geographic levels, not just national or other high-level aggregates. Phraseology that we invoked above is applicable here: analysis is a key first step, no matter how imprecise the source data might be or how caveated results must be. Small-scale field collection of GPS readings and independent listings may not generalize well, but modeling and small-area estimation approaches could usefully be introduced; perhaps a spatial data quality estimate for every small county is infeasible, but an estimate for "places like us" (collections of places that are similar by demographic characteristics or other stratification variables) could still usefully steer geographic updating resources. An earlier National Research Council (2009:119–128) report discussed a framework for modeling census quality using both MAF/TIGER and census operational data as inputs, and that work may suggest possible directions.

Related to another core research area, another priority for geographic work is to prepare for the possible use of administrative records data in geographic update operations. In addition to a person-level data file, the Bureau's current StARS system also generates a listing of addresses, dubbed the Master Housing File (MHF). Just as current work with StARS on the person-level side has largely been limited to looking at gross counts, so too has the utility of the MHF as an update source for—or quality check of—the MAF/TIGER databases been largely unexplored to date. Bureau staff attempt to use the TIGER database to geocode the MHF—associate each address with a specific geographic code—but to date has not delved deeply into the attributes of MHF addresses that do or do not geocode. Likewise, the year-to-year flux in MHF content—"births" and "deaths" of addresses—remains to be explored.

It is our understanding that the Census Bureau is working on converting the samples for its ongoing demographic surveys to use the MAF as their address source, much like the ACS does now. The status quo for the current surveys is to draw their sample from parts of four different address frames—an address frame (separate from the MAF), an area frame, a GQ inventory, and a listing of new construction addresses. Switching the surveys to use the MAF as a base has the advantage of making the MAF a fuller "corporate resource" within the Census Bureau; it is also useful in that it gives the current surveys a direct stake in the quality of MAF/TIGER, and so could facilitate the use of survey interviewers as part of regular geographic quality assessment (as mentioned above). Our charge is focused on the decennial census and its specific operations, but we think it entirely appropriate to support the use of the MAF in all of the regular current surveys; updates and improvements to MAF/TIGER based on regular use of those systems ultimately accrue to the quality of the census.

References

Côté, A.-M. and D. Laroche (2009). The Internet: A new collection method for the census. In *Proceedings of Statistics Canada Symposium 2008—Data Collection: Challenges, Achievements and New Directions*. Ottawa: Statistics Canada. Available: http://www.statcan.gc.ca/pub/11-522-x/2008000/article/10986-eng.pdf [accessed December 17, 2010].

Groves, R. M. (2009, September 22). *2010 Census—Operational Assessment*. Prepared statement before the Subcommittee on Information Policy, Census, and National Archives, Committee on Oversight and Government Reform, U.S. House of Representatives: U.S. Census Bureau.

Holmes, A. (2008, January 2). Census program to use handheld computers said to be in 'serious trouble'. *Government Executive.com*. Available: http://www.govexec.com/dailyfed/0108/010208h1.htm [accessed November 16, 2010].

Judson, D. H. and B. Bye (2003, October 21). *Synthesis of Results from the Administrative Records Experiment in 2000 (AREX 2000)*. Washington, DC: U.S. Census Bureau.

Kincannon, C. L. (2006, June 6). The 2010 decennial census program. Prepared statement, testimony before the Subcommittee on Federal Financial Management, Government Information, and International Security of the Committee on Homeland Security and Governmental Affairs, U.S. Senate.

MITRE Corporation (2007, January 17). An assessment of the risks, costs, and benefits of including the Internet as a response option in the 2010 decennial census. Version 2.0. Report prepared for the U.S. Census Bureau.

National Research Council (1995). *Modernizing the U.S. Census*. Panel on Census Requirements in the Year 2000 and Beyond, Barry Edmonston and Charles Schultze, eds., Committee on National Statistics, Commission of Behavioral and Social Sciences and Education. Washington, DC: National Academy Press.

National Research Council (2004a). *The 2000 Census: Counting Under Adversity*. Panel to Review the 2000 Census, Constance F. Citro, Daniel L. Cork, and Janet L. Norwood, eds., Committee on National Statistics. Washington, DC: The National Academies Press.

National Research Council (2004b). *Reengineering the 2010 Census: Risks and Challenges*. Panel on Research on Future Census Methods, Daniel L. Cork, Michael L. Cohen, and Benjamin F. King, eds., Committee on National Statistics, Division of Behavioral and Social Sciences and Education. Washington, DC: The National Academies Press.

National Research Council (2009). *Coverage Measurement in the 2010 Census*. Panel on Correlation Bias and Coverage Measurement in the 2010 Decennial Census, Robert M. Bell and Michael L. Cohen, eds., Committee on National Statistics, Division of Behavioral and Social Sciences and Education. Washington, DC: The National Academies Press.

National Research Council (2010). *Envisioning the 2020 Census*. Panel on the Design of the 2010 Census Program of Evaluations and Experiments, Lawrence D. Brown, Michael L. Cohen, Daniel L. Cork, and Constance F. Citro, eds. Committee on National Statistics, Division of Behavioral and Social Sciences and Education. Washington, DC: The National Academies Press.

Statistics Canada (2008). *Earnings and Incomes of Canadians Over the Past Quarter Century, 2006 Census*. Catalogue no. 97-563-X. Ottawa: Statistics Canada.

U.S. Census Bureau (1976). *Procedural History PHC(R)-1, 1970 Census of Population and Housing*. Washington, DC: U.S. Government Printing Office.

U.S. Census Bureau (2010, February). *U.S. Census Bureau's Budget Estimates, As Presented to Congress, February 2010: Fiscal Year 2011*. Washington, DC: U.S. Census Bureau.

U.S. Department of Commerce, Office of Inspector General (2006, March). *U.S. Census Bureau—Census 2010: Revised Field Data Collection Automation Contract Incorporated OIG Recommendations, But Concerns Remain Over Fee Awarded During Negotiations*. Final Report No. CAR-18702. Washington, DC: U.S. Department of Commerce.

U.S. Government Accountability Office (2007, July 17). *2010 Census: Preparations for 2010 Census Underway, But Continued Oversight and Risk Management Are Critical*. Report GAO-07-1106T. Testimony of Mathew J. Scirè and David A. Powner before the Subcommittee on Federal Financial Management, Government Information, Federal Services and International Security, Committee on Homeland Security and Government Affairs, U.S. Senate. Washington, DC: U.S. Government Printing Office.

U.S. Government Accountability Office (2008, July). *2010 Census: Census Bureau's Decision to Continue with Handheld Computers for Address Canvassing Makes*

Planning and Testing Critical. Report GAO-08-936. Washington, DC: U.S. Government Printing Office.

U.S. Office of Management and Budget (2010). Strengthening federal statistics. In *Analytical Perspectives, Budget of the United States Government, Fiscal Year 2011*, Chapter 18, pp. 315–319. Washington, DC: U.S. Government Printing Office.

– A –

Charge of the Panel to Review the 2010 Census

[The Panel to Review the 2010 Census will] conduct a comprehensive evaluation of the statistical methods and operational procedures for the 2010 census. The panel will address, in particular, methods and procedures that may affect the completeness and quality of the census enumeration, including preparation of the Master Address File and associated spatial data, census operations affecting group quarters enumeration, housing unit enumeration, and completeness of census coverage, the Census Coverage Measurement Program's field, matching, and estimation activities, use of technology, and management of the 2010 census. The panel will not only evaluate the 2010 census as such, but also draw lessons for design and planning for a more cost-effective 2020 census. The panel will issue a final report at the completion of a five-year study with its findings and recommendations and will issue one or more interim reports as needed to address particular topics for which it is important to provide an early assessment of 2010 operations and advice for 2020 census planning.

– B –

Site Visits by Panel Members and Staff

Location	Date (2010)	Participants
General and Early Census Operations		
DC West, DC (LCO 2313)	February 19	Warren Brown, Daniel Cork, Anthony Mann
Culver City, CA (LCO 3214)	February 22	Art Geoffrion, Judith Seltzer, Daniel Cork, Anthony Mann
Los Angeles RCC	February 22	Art Geoffrion, Judith Seltzer, Daniel Cork, Anthony Mann
Dallas Central, TX (ELCO 3034)	February 23	Thomas Cook, Daniel Cork, Anthony Mann
Dallas RCC	February 23	Thomas Cook, Daniel Cork, Anthony Mann
Concord, NH (ELCO 2131)	March 12	Donald Cooke, Susan Hanson, Daniel Cork
Billings, MT (ELCO 3129)	March 18	Daniel Cork
New Orleans, LA (LCO 3018)	March 18	Michael Cohen, Anthony Mann
Albuquerque, NM (ELCO 3139)	March 19	Jack Baker, Daniel Cork, Anthony Mann
Nonresponse Follow-up (NRFU) Preparation and Early NRFU		
San Francisco West, CA (LCO 2723)	March 30	Matthew Snipp, Daniel Cork, Anthony Mann
Louisville, KY (ELCO 2816)	April 6	Warren Brown, Donald Cooke, Constance Citro, Daniel Cork, Anthony Mann

Phoenix Central, AZ (ELCO 3112)	April 9	Matthew Snipp, Daniel Cork
Dallas RCC	April 16	Matthew Snipp, Constance Citro
Milwaukee, WI (ELCO 2546)	April 16	Daniel Cork
Atlanta RCC	April 19	Warren Brown, Daniel Cork, Anthony Mann
Athens, GA (LCO 2953)	April 20	Warren Brown, Daniel Cork, Anthony Mann
Austin, TX (LCO 3027)	April 21	Daniel Cork, Anthony Mann
Palo Alto, CA (LCO 2717)	April 22	Matthew Snipp
Fairfax, VA (LCO 2855)	April 23	Daniel Cork, Anthony Mann
North Las Vegas, NV (LCO 3137)	April 30	Donald Cooke, Daniel Cork
NRFU		
Kansas City RCC	May 3	Daniel Cork, Anthony Mann
Kansas City, KS (LCO 2621)	May 3	Daniel Cork, Anthony Mann
Savannah, GA (ELCO 2965)	May 7	Warren Brown
Joliet, IL (LCO 2531)	May 14	Daniel Cork
Rockville, MD (LCO 2319)	May 19	Michael Larsen, Nathaniel Schenker, Daniel Cork, Michael Cohen, Anthony Mann
Philadelphia RCC	May 24	Daniel Cork, Anthony Mann
Philadelphia West, PA (LCO 2342)	May 24	Daniel Cork, Anthony Mann
Queens Northeast/Flushing, NY (LCO 2234)	May 24	Daniel Cork, Anthony Mann
Burlington, VT (ELCO 2146)	May 28	Donald Cooke, Susan Hanson, Daniel Cork, Anthony Mann
Santa Fe, NM (LCO 3141)	June 1–2	Jack Baker, Daniel Cork
Ventura, CA (LCO 3244)	June 7	Daniel Cork, Anthony Mann
El Paso, TX (LCO 3039)	June 8	Daniel Cork, Anthony Mann
Columbus Central, OH (LCO 2436)	June 18	Ivan Fellegi, Daniel Cork
Homestead, FL (LCO 2932)	June 22	Daniel Cork, Anthony Mann
Durham, NC (LCO 2823)	June 23	Daniel Cork, Anthony Mann
Inglewood, CA (LCO 3225)	June 23	Donald Cooke, Art Geoffrion, Judith Seltzer
Late Census Operations and Census Coverage Measurement		
Wichita, KS (LCO 2623)	June 29	Daniel Cork
Chicago Near South, IL (ELCO 2518)	June 30	John Thompson, Daniel Cork
Anchorage, AK (ELCO 2711)	July 26	Daniel Cork
Seattle RCC	July 28	Daniel Cork
Duluth, MN (ELCO 2625)	August 5	Daniel Cork
New York RCC	August 11	Daniel Cork
Bronx Northwest, NY (LCO 2220)	August 12	Ivan Fellegi, Daniel Cork
Cincinnati Suburban, OH (LCO 2431)	August 17	Roger Tourangeau (CNSTAT), Daniel Cork
Jacksonville North, FL (LCO 2933)	August 18	Daniel Cork
Boston RCC	August 23	Donald Cooke, Daniel Cork

Charleston, WV (ELCO 2446)	August 26	Daniel Cork
Philadelphia RCC	August 30	Michael Cohen, Daniel Cork
Detroit RCC	November 4	Daniel Cork
Paper Data Capture and Telephony Operations		
Essex, MD, PDCC	March 31	John Thompson
Jeffersonville, IN, NPC	April 6	Warren Brown, Donald Cooke, Constance Citro, Daniel Cork, Anthony Mann
Essex, MD, PDCC	April 19	Susan Hanson, Michael Cohen, Anthony Mann
Phoenix, AZ, PDCC	April 19	Matthew Snipp, Daniel Cork
Essex, MD, PDCC	April 27	Thomas Cook, Donald Cooke, George Ligler, Nathaniel Schenker, Daniel Cork, Anthony Mann
Greenbelt, MD, DRIS Command Center	July 9	Roger Tourangeau (CNSTAT), Daniel Cork, Anthony Mann
Lawrence, KS, Call Center	July 13	Daniel Cork, Anthony Mann
Phoenix, AZ, Call Center	July 22	Matthew Snipp, Daniel Cork, Anthony Mann
Phoenix, AZ, PDCC	July 22	Matthew Snipp, Daniel Cork, Anthony Mann

NOTES: CNSTAT, Committee on National Statistics (member); DRIS, Decennial Response Integration System; ELCO, Early Local Census Office; LCO, Local Census Office; NPC, National Processing Center; PDCC, Paper Data Capture Center; RCC, Regional Census Center. ELCOs opened to support the Address Canvassing operation in early 2009, dividing their jurisdictions across one or more later-opening regular LCOs when the full set of 494 LCOs opened in late 2009. The first two digits of each ELCO or LCO number denote the parent census region; for example, the 27 prefix indicates offices in the Seattle region.

– C –

Biographical Sketches of Panel Members and Staff

Thomas M. Cook (*Co-Chair*) is former president of SABRE Decision Technologies, where he was responsible for a 2,700-person consulting and software development company specializing in providing solutions to the travel and transportation industry. He has served as chairman and CEO of CALEB Technologies Corporation, president of T.C.I. Consulting, and senior counselor at McKinsey and Company. In his career at AMR Corporation, he was director of operations research for American Airlines, from which SABRE emerged as a separate entity. He has also held positions at the University of Tulsa and Arthur Young and Company. He was elected to the National Academy of Engineering in 1995 for leadership in advancing operations research within the transportation industry and has served as president of the Institute of Management Sciences and the Institute of Operations Research and the Management Sciences (INFORMS). He holds a master's degree in business administration from Southern Methodist University and a Ph.D. in operations research from the University of Texas.

Janet L. Norwood (*Co-Chair*) served as U.S. Commissioner of Labor Statistics from 1979 to 1992. She has served as a senior fellow at the Urban Institute, a director and vice-chair of the board of the National Opinion Research Center (NORC) at the University of Chicago, and as counselor and senior fellow at the Conference Board. At the National Research Council, she has served on numerous study panels and chaired the Panel to Review the 2000 Census and the Panel on Statistical Programs of the Bureau of

Transportation Statistics (BTS); she is also a past member of the Committee on National Statistics and the Division of Engineering and Physical Sciences. She is a member of advisory committees at the National Science Foundation, at several statistical agencies, and at universities, and has chaired the advisory committee for BTS. She has received honorary L.L.D. degrees from Carnegie Mellon, Florida International, Harvard, and Rutgers Universities. She is a fellow and past president of the American Statistical Association, a member and past vice president of the International Statistical Institute, an honorary fellow of the Royal Statistical Society, and a fellow of the National Academy of Public Administration and the National Association of Business Economists. She has a B.A. degree from Rutgers University and M.A. and Ph.D. degrees from the Fletcher School of Law and Diplomacy of Tufts University.

Jack Baker is senior research scientist in the Geospatial and Population Studies Program at the University of New Mexico. Since 2006, he has represented New Mexico in the Federal-State Cooperative Programs on Population Estimates (FSCPE) and Population Projections (FSCPP). He participated extensively in preparations for the 2010 census, with an emphasis on Master Address File improvement efforts including the 2010 Local Update of Census Addresses (LUCA) program and the 2010 Count Review Program (chairing the FSCPE committee that focused on redesigning this process and as a consultant to the Bureau on use of GIS technology to perform the review). He continues to serve on numerous FSCPE committees. His scientific research focuses primarily on methods for modeling small area populations using incomplete data, geospatial demographic methods, historical demography, and biodemography. He received a B.A. degree from the University of North Dakota and M.S. and Ph.D. degrees from the University of New Mexico, all in anthropology.

Warren Brown is senior public service associate and director of the applied demography program at the Carl Vinson Institute of Government at the University of Georgia; the applied demographic program is charged with providing demographic population estimates and projections for the state of Georgia, in partnership with the state Office of Planning and Budget. Previously, he was director of the Program on Applied Demographics at Cornell University, in which capacity he was responsible for producing population estimates and projections for the state of New York. He has represented New York in the Census Bureau's Federal-State Cooperative Programs for Population Estimates and Population Projections, serving as chair of the population estimates group. He also served on the Population Association of America's Committee on Challenges to Population Estimates, Advisory Committee on the Demographic Full Count Review, and Committee on

Applied Demography. At Cornell, he also served as research director of the university's Census Research Data Center. He received a B.A. in religious studies from the University of Virginia, an M.A. in sociology from the New School for Social Research, and a Ph.D. in development sociology from Cornell University.

Donald Cooke is community mapping evangelist at Esri in Redlands, California. He was a member of the 1967 Census Bureau team that developed the Dual Independent Map Encoding (DIME) topological approach to a spatial database as part of the New Haven Census Use Project. The DIME methodology was a key predecessor to the Census Bureau's Topologically Integrated Geographic Encoding and Referencing (TIGER) system and of the modern geographic information systems industry. In 1980 he founded Geographic Data Technology, Inc. (GDT), with which the Census Bureau contracted to digitize the original TIGER data files. GDT was acquired by Tele Atlas in 2004, and he was chief scientist at Tele Atlas North America through February 2010. He received the Urban and Regional Information Systems Association's Horwood award in 2004 and Esri's lifetime achievement award in 2007. At the National Research Council, he has served on the Mapping Science Committee. He is a graduate of Yale University and studied civil engineering systems at the Massachusetts Institute of Technology.

Daniel L. Cork *(study director)* is a senior program officer for the Committee on National Statistics, currently serving as study director of the Panel to Review the 2010 Census. He joined the CNSTAT staff in 2000 and has served as study director or program officer for several census panels, including the Panels on Residence Rules in the Decennial Census, Research on Future Census Methods (2010 Planning panel), and Review of the 2000 Census. He also directed the Panel to Review the Programs of the Bureau of Justice Statistics (in cooperation with the Committee on Law and Justice) and was senior program officer for the Panel on the Feasibility, Accuracy, and Technical Capability of a National Ballistics Database (joint with the Committee on Law and Justice and the National Materials Advisory Board). His research interests include quantitative criminology, geographical analysis, Bayesian statistics, and statistics in sports. He has a B.S. in statistics from George Washington University and an M.S. in statistics and a joint Ph.D. in statistics and public policy from Carnegie Mellon University.

Ivan P. Fellegi is chief statistician emeritus of Canada, having served as chief statistician from 1985 to 2008. He joined Statistics Canada (then the Dominion Bureau of Statistics) in 1957, serving as director of sampling research and consultation and director general of methodology and systems, assistant chief statistician, and deputy chief statistician before his appointment as chief statistician. He has published extensively in the areas of census and survey methodology, in particular on consistent editing rules and record linkage. A past chair of the Conference of European Statisticians of the United Nations Economic Commission for Europe, he is an honorary member and past president of the International Statistical Institute, an honorary fellow of the Royal Statistical Society, past president of the International Association of Survey Statisticians, and past president and Gold Medal recipient of the Statistical Society of Canada. He was made Member of the Order of Canada in 1992 and promoted to Officer in 1998 and has received the nation's Outstanding Achievement Award; he has also provided advice on statistical matters to his native Hungary following its transition to democracy and, in 2004, was awarded the Order of Merit of the Republic of Hungary. At the National Research Council, he was a member of the Panel on Privacy and Confidentiality as Factors in Survey Response, the Panel on Census Requirements in the Year 2000 and Beyond, the Panel on Decennial Census Methodology, and the Panel on the Design of the 2010 Census Program of Evaluations and Experiments. He has a B.Sc. from the University of Budapest and M.Sc. and Ph.D. degrees in survey methodology from Carleton University.

Arthur M. Geoffrion is James A. Collins professor of management emeritus (recalled) at the University of California, Los Angeles, Anderson School of Management. The author of more than 60 published works ranging from mathematical programming to the implications of the digital economy for management science, he has consulted extensively on applications of optimization to problems of distribution, production, and capital budgeting. In 1978 he co-founded INSIGHT, Inc., a management consulting firm specializing in optimization-based applications in supply-chain management and production. In 1982, he founded what is now the Management Science Roundtable, an organization composed of the leaders of operations research groups in 50–60 companies. His editorial service includes eight years as department editor (mathematical programming and Networks) of Management Science. He has served as the president of the Institute of Management Sciences and received that institute's distinguished service medal; he is also a fellow and past president of the Institute for Operations Research and the Management Sciences and recipient of its George E. Kimball Medal. He is an elected member of the National Academy of Engineering. He received his B.M.E. and his M.I.E. at Cornell University, and his Ph.D. at Stanford University.

Susan Hanson is research professor in the Graduate School of Geography at Clark University and has previously served as the school's director. An urban geographer, her current research focuses on understanding how gender, geographic opportunity structures, and geographic rootedness affect entrepreneurship in cities, as well as on understanding the emergence of sustainable versus unsustainable practices in urban areas. Prior to joining the Clark faculty, she held faculty appointments at Middlebury College and the State University of New York at Buffalo. She has served as editor of Economic Geography, the Annals of the Association of American Geographers, Urban Geography, and The Professional Geographer, and has been on the editorial boards of numerous other journals. She was elected to the National Academy of Sciences in 2000; she is a past president of and was awarded lifetime achievement honors by the Association of American Geographers and is a fellow of the American Association for the Advancement of Science. She has served on several National Research Council panels and committees, including the Committee on National Statistics' Panel on Measuring Business Formation, Dynamics, and Performance. She has a B.A. in geography from Middlebury College and a Ph.D. in geography from Northwestern University.

Michael D. Larsen is associate professor of statistics and member of the faculty of the Biostatistics Center at George Washington University. Previously, he was associate professor of statistics at Iowa State University. He has served as executive editor of *CHANCE* magazine, and as associate editor of the *Annals of Applied Statistics*, the *Journal of Statistics Education*, and the *Journal of Official Statistics*. He has served on the Census Advisory Committee of the American Statistical Association. He received his B.A. in mathematics and his M.A. and Ph.D. degrees in statistics from Harvard University.

George T. Ligler is a private consultant in Potomac, Maryland. He has extensive experience in information management and software and computer system engineering, as is evident from his work at Burroughs Corporation (1980–1982), Computer Sciences Corporation (1984–1988), and at GTL Associates, a private company that he founded. At the National Research Council, he served on the Computer Science and Telecommunications Board's Committee to Review the Tax Systems Modernization of the Internal Revenue Service. He also served as a member of the Panel on Research on Future Census Methods (2010 census planning), and was a member of the expert committee separately formed by the U.S. Secretary of Commerce to advise on options for the Census Bureau's replan of its Field Data Collection Automation Contract in early 2008. A Rhodes scholar, he received his B.S. in mathematics from Furman University in 1971 and his M.Sc. and D.Phil. from Oxford University.

Nathaniel Schenker is associate director for research and methodology at the National Center for Health Statistics, having previously served as senior scientist. He is also adjunct professor in the Joint Program in Survey Methodology administered by the University of Maryland, University of Michigan, and Westat. Prior to that he was associate professor in the Department of Biostatistics at the University of California, Los Angeles, and prior to that he was a mathematical statistician at the Bureau of the Census. His research interests include handling incomplete data, census and survey methods, survival analysis, statistical computation, and applications of statistics to the health and social sciences. He is a past vice president and past board member of the American Statistical Association. He also served as program chair of the Joint Statistical Meetings, and he was editor of a special section of the *Journal of the American Statistical Association* entitled "Undercount in the 1990 Census." At the National Research Council, he was a member of the Panel on Alternative Census Methodologies. He received the Roger Herriot Award for Innovation in Federal Statistics from the American Statistical Association, he is a fellow of the American Statistical Association, and he is an elected member of the International Statistical Institute. He received his A.B. in statistics from Princeton, and his S.M. and Ph.D. in statistics from the University of Chicago.

Judith A. Seltzer is professor of sociology at the University of California, Los Angeles. Previously, she was on the faculty of the University of Wisconsin–Madison, where she contributed to the development and implementation of the National Survey of Families and Households. Her research interests include kinship patterns, intergenerational obligations, relationships between nonresident fathers and children, and how legal institutions and other policies affect family change. She was part of a cross-university consortium to develop new models for explaining family change and variation and a member of the design team for the Los Angeles Family and Neighborhood Survey. At the National Research Council, she has served on the Panel on Residence Rules in the Decennial Census and the Panel on the Design of the 2010 Census Program of Evaluations and Experiments. She has master's and Ph.D. degrees in sociology from the University of Michigan.

C. Matthew Snipp is Burnet C. and Milfred Finley Wohlford professor of sociology at Stanford University. At Stanford, he is currently serving as director of the Secure Data Center of the Institute for Research in the Social Sciences and of the Center for Comparative Studies of Race and Ethnicity; he is also on the faculty of the Native American Studies program. He has written extensively on American Indians and has written specifically on the interaction of American Indians and the U.S. census. He has served on the Census Bureau's Technical Advisory Committee on Racial and Ethnic

Statistics and the Native American Population Advisory Committee. Prior to moving to Stanford, he was associate professor and professor of rural sociology at the University of Wisconsin–Madison, where he held affiliate appointments with several other units, and assistant and associate professor of sociology at the University of Maryland. At the National Research Council, he was a member of the Panel on Research on Future Census Methods (2010 census planning) and the Panel on Residence Rules in the Decennial Census. He received his A.B. in sociology from the University of California, Davis, and his M.S. and Ph.D. in sociology from the University of Wisconsin–Madison.

John H. Thompson is president of the National Opinion Research Center (NORC) at the University of Chicago. Prior to his appointment as president, he was executive vice president for survey operations, in which capacity he provided oversight and direction for NORC's Economics, Labor Force, and Demography Research Department and the Statistics and Methodology Department. He also served as project director for the National Immunization Survey, conducted on behalf of the Centers for Disease Control and Prevention from November 2004 through July 2006. He joined NORC following a 27-year career at the U.S. Census Bureau, culminating in service as principal associate director for programs. As associate director for decennial census (1997–2001) and chief of the Decennial Management Division (1995–1997), he was the chief operating officer of the 2000 census, overseeing all aspects of census operations. In this capacity, he also chaired the Bureau's Executive Steering Committee for Accuracy and Coverage Evaluation Policy, an internal working group tasked to provide guidance to the director of the Census Bureau and the secretary of commerce concerning statistical adjustment of 2000 census figures. He has received a Presidential Rank Award of Meritorious Executive and Gold, Silver, and Bronze Medals from the U.S. Department of Commerce. At the NRC, he served on the Panel on the Design of the 2010 Census Program of Evaluations and Experiments. He is a fellow of the American Statistical Association. He has bachelor's and master's degrees in mathematics from Virginia Polytechnic Institute and State University.

COMMITTEE ON NATIONAL STATISTICS

The Committee on National Statistics was established in 1972 at the National Academies to improve the statistical methods and information on which public policy decisions are based. The committee carries out studies, workshops, and other activities to foster better measures and fuller understanding of the economy, the environment, public health, crime, education, immigration, poverty, welfare, and other public policy issues. It also evaluates ongoing statistical programs and tracks the statistical policy and coordinating activities of the federal government, serving a unique role at the intersection of statistics and public policy. The committee's work is supported by a consortium of federal agencies through a National Science Foundation grant.